Genetic Laws Governing
The Breeding of Standard Fowls

Outbreeding, Inbreeding and Linebreeding All Recognized Breeds of Poultry

by Wetherell Henry Card

with an introduction by Jackson Chambers

This work contains material that was originally published in 1912.

This publication is within the Public Domain.

*This edition is reprinted for educational purposes
and in accordance with all applicable Federal Laws.*

Introduction Copyright 2018 by Jackson Chambers

The World's Largest Selection of Vintage Poultry Books

www.VintagePoultry.com

Self Reliance Books

Get more historic titles on animal and stock breeding, gardening and old fashioned skills by visiting us at:

http://selfreliancebooks.blogspot.com/

Introduction

I am pleased to present yet another title on Poultry.

The work is in the Public Domain and is re-printed here in accordance with Federal Laws.

As with all reprinted books of this age that are intended to perfectly reproduce the original edition, considerable pains and effort had to be undertaken to correct fading and sometimes outright damage to existing proofs of this title. At times, this task is quite monumental, requiring an almost total "rebuilding" of some pages from digital proofs of multiple copies. Despite this, imperfections still sometimes exist in the final proof and may detract from the visual appearance of the text.

I hope you enjoy reading this book as much as I enjoyed making it available to readers again.

Jackson Chambers

FOREWORD

THE illustrations of White Laced Red Cornish fowls are placed in this book because in the origin and evolution of this breed I learned the prime truths and fundamental principles of animal breeding as well as the superiority and prepotency of all animals line-bred and that like begets like when these principles and laws are followed. The first illustration is the original male on which this breed was founded. He was sired by a White Cornish; his dam was one quarter Brahma, one quarter Shamo Jap. Game and one half Dark Cornish; he was bred back for five consecutive years to his own pullets with the results of establishing color and markings identical with his own. Hence, color and markings and characteristics for a breed were established which has been bred and perpetuated in line ever since as per chart. ¶ The history of all animal breeding where racial characteristics are desired, is grand and all sufficient proof that these principles are eternally right. Pure bred domestic animals of any race or kind are evolutions brought to high standards of perfection according to kind by a strict adherence to line-breeding and nature's laws. Thus, the laws governing the breeding of fancy fowls as laid down in these pages are simply plain statements of plain everlasting truths, proven and reproven in all animal life. The results of negligence or ignorance are only further proofs of these great laws.

THE AUTHOR.

LAWS GOVERNING THE BREEDING OF STANDARD FOWLS

> O, what men dare do! What men may do! What men daily do, not knowing what they do!
> —From "Much Ado About Nothing."

CHAPTER I

KNOWLEDGE is power. While there are many things still unsolved and many perhaps unsolvable, the greatest knowledge of all is to recognize the fact of the inevitableness, the immutability, the irrevocableness of certain laws which govern every phase of nature as regards a reproduction of kind.

Right at the beginning I wish to state that this book contains no theories nor conjectures, but simple facts regarding these certain laws, which I have learned, observed and proved in forty years of breeding fancy poultry. I want my readers to notice in particular how applicable is the thought that twice two is four in any language; to notice how the many seeming intricacies and unsolvable puzzles of fancy poultry breeding will resolve themselves into plain simple logic and law which in its application is as true and as accurate, aye, and as inevitable as the turning of the earth on its axis; that the intricate and puzzling things are not parts of the law so much as they are the results of broken laws.

Whether under civil, financial, physical, natural, spiritual or mental law, the man who obeys

is free of the law. The transgressor suffers the restrictions, the complications, the punishments and ultimate failures. Ignorance is no excuse but instead is the "bar sinister;" therefore I want my readers to take careful note of the fact that to succeed one must understand the law which came to the "Old Vets" through hard knocks against real facts, and not dreams or theories.

Just the same as twice two is four in any language so are the fundamental laws of nature the same in any breed, kind, species or variety in the animal kingdom. When one comes to a knowledge of this fact the mists of uncertainty disappear and the sun of absolute confidence shines. This comes not in a day nor a year, but from years of puzzling, proving and investigating. The breeding of all domestic animals comes under

the very same law, whether dogs, sheep, horses or poultry. Notice, I say domestic animals, and I emphasize this because domestic animals in nearly every instance are mongrels or evolutions from amalgamations of several bloods of the same species, therefore under natural law radically different in most respects (especially the first prime requisite, in-breeding) from the natural law which governs wild animals, which are pure of their kind and not amalgamations.

The law of atavism in domestic animals is the indicator which keeps tabs on the different bloods, marks and labels them according to kind and flaunts the warning signal at the opportune time to the observant progressive breeder who obeys the law.

In wild animals like begets like unerringly and they are immune to the laws of in-breeding and atavism because they are the complete personification of atavism, or reverting back to kind.

Now, dear readers, don't make the mistake of thinking that the laws which govern wild and domestic animals are opposites; the difference lies only in the impurity of the domestic against the purity of the wild. The radical point of difference is the method necessary to bring domestic animals to a somewhat near state of purity. In spite of man's egotism and pride in his handiwork, atavism is the one fact that proves his work mongrelism, therefore imperfect and it can never be otherwise, the degrees varying according to the care and knowledge of the breeder. The inevitable law of nature will ever make itself apparent whenever the fusing of two or more bloods occurs. For example, the white face found occasionally in Leghorns shows the White Faced Spanish infusion; the persistent stub on the shanks and feet of the clean legged varieties labels the drop of Asiatic blood; the side sprig on the single comb, the taint of the rose comb; occasional colored feathers in black or white birds mean that at some time in their history colored ancestors were in their line; and many other characteristics more apparent such as shape, and even disposition. All this proves this great law as one to be reckoned with by the successful breeder of domestic animals. Notice that along this line I am persisting in writing "domestic animals" instead of domestic fowls, because I wish to impress the Rhode Island Red breeder, as well as the Cochin breeder; the Bantam breeder equally with the Plymouth Rock breeder, that nature's laws hold good and are absolute and equally applicable in one breed as in the other. With this thought well established and fixed, the intricacies of breeding are no longer the bugbears of the fancier and breeder.

What men daily do, aye yearly do, not knowing what they do, strikingly illustrates haphazard, blind breeding of fancy fowls the world over. Haphazard is the gambler's chance throw; investigation, study and method mean knowledge and surety. Haphazard is a paint-filled brush thrown at random on a canvas with an ugly blot the result. When reason and intelligent deductions direct the brush a beautiful picture appears. One proof of the generality of haphazard as related to fancy fowls is the frequent use of the word "phenomenon." Webster defines the word as an appearance whose cause is not immediately obvious. A phenomenon in the show pen is

LAWS GOVERNING THE BREEDING

hardly ever reproduced by the ordinary manner of breeding fowls; because the reason of its appearance is not obvious; this explains why the breeding of such a bird generally produces a majority of culls. The use of the word phenomenon seems to be a tacit admission of ignorance, yet a word behind which that ignorance may hide. Knowledge produces a majority of specimens better than their kind; haphazard may produce and may not produce phenomena of obscure origin; one is certainty, the other chance.

In-breeding by judicious line-breeding is a peculiar anomaly in the law of nature, because there is an apparent transgression with an absolute abeyance; yet it is only the unnatural against the natural, purity warring with impurity, which demands laws to fit the case for a successful culmination.

In the breeding of domestic animals to a high state of perfection according to prescribed standards, inbreeding is the chief factor or law. (Inbreeding means breeding from close relationship.) Yet according to the natural laws in breeding domestic animals it tends to a certain deterioration; therefore the careful thinking breeder adds the word judicious to inbreeding which, defined in combination, means line breeding and when thoroughly understood means success. Judicious inbreeding has laws of its own and "obey" is the open sesame.

The first law of judicious inbreeding is stamina and vigor and is in perfect accordance with the law of all nature; i. e., survival of the fittest. The term judicious in this instance means not breeding too close nor for too long a time, but close and long enough to fix type and color according to standard, at the same time estab-

lishing lines of the same blood from which to feed the main line in progress and then, when type and color are well established, to breed from the farthest removed in the same blood lines, never forgetting stamina, thus forestalling or keeping in abeyance that result of the transgression of natural laws among domestic animals,

LAWS GOVERNING THE BREEDING

"the almost inevitable deterioration" caused by inbreeding. Thus is there transgression almost in harmony with obedience. Among the animals of the wild, nature exacts no such paradoxical conditions and can offer no help to the beginner in the study of in and line breeding.

The laws of Mendel, Galton and Darwin are too highly scientific for the average breeder. The many ramifications of the recessive and dominant are dark and devious ways which lead the beginner or the mind not scientifically trained into a wilderness of conjecture and the morass of perplexity, confounding and confusing all efforts into an aimless circle which ends only in failure. Throughout all these pages it is my purpose to avoid all semblance of thought, methods or laws which the ordinary breeder and average fancier cannot easily understand and apply for his betterment. Mendelism may be the true solution of the breeding problem and can perhaps be successfully used and applied with far reaching results by one scientific enough to understand its premises. For the average fancier to delve into the scientific mysteries of Mendelism is like chasing the rainbow for the pot of gold.

By using common sense reasoning in the study and application of nature's laws, with wide open eyes, as cause and effect unroll in plain and understandable formation in the reproduction and evolution of all animal life, the diamonds of a material success are unearthed. Therefore we present a surer and safer method for the average breeder or fancier, because devoid of technical terms and entangling ideas and premises, so confusing and obscure to the common man.

Be it far from me to decry the value of Mendelism or any deep scientific delving into the mysteries of life by men who devote their lives to an earnest, honest search after the truth. I simply state a fact when I say that these higher sciences

are as an unknown language of high sounding words which convey no meaning to the average breeder. It is also a fact that the majority of successful breeders the world over are those who have built and bred on the safe ground of common sense reasoning as applied to animal breeding; a sort of a science which produces, protects and preserves by its very simplicity. To bring this thought of judicious inbreeding out more in detail

LAWS GOVERNING THE BREEDING

I shall explain the method of using the system which dates back since man first raised fancy chickens. Sebright bantams first brought it to my notice and after a prolonged trial with Polish fowls to a success I concluded it was applicable to any breed and so it has proved.

First and foremost, dear reader, to make the right start, buy the very best male and female possible for your line-breeding. This pair are your ideals to work from and the best is only good enough for such a start, and then again, contrary to popular belief, you should breed from an ideal as well as up to an ideal. The first year's results from your ideals contain half the blood of the sire and half the blood of the dam. (See chart on page 15.) Now the next step is to take the best pullet from this mating and breed her to her sire and the best cockerel should be bred to his dam. For convenience sake we will say this is 1911 mating. Next take pullet from 1911 mating and use again on first sire; this pullet being granddaughter of her sire and containing three quarters of his blood. Take cockerel from the mating of 1911 with dam and do likewise; this cockerel now carries three quarters of the blood of dam. The next mating carries us to 1912. The young from these matings carry seven-eighths of the blood of original sire and dam respectively.

To protect the line from the effects of too close inbreeding one should no longer breed back to original sire or dam. The 1912 mating back to original sire contains only one-eighth of the blood

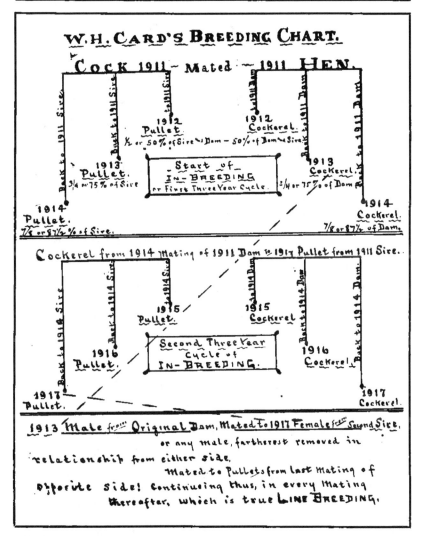

of the original dam. The 1912 mating back to the original dam contains only one-eighth of the blood of the original sire. A line is now started; therefore to keep within the blood of the original sire and dam take the best cockerel from the dam side and mate to the best pullet from the sire side; this mating carries us to 1913 results. Take best pullet from this mating and mate to her sire; take best cockerel from this mating and mate to his dam. Return pullet of each year of 1914 and 1915 to same sire; and return cockerel each year 1914 and 1915 to same dam.

The line is now so well established that close inbreeding is no longer necessary, so in order to enhance stamina and to firmly fix it in the line go back to 1912 mating for a male to use on one line and a male from 1913 mating to use on the other line. Always keep accurate account of every mating. Henceforth use males from the farthest removed in relationship from these lines, and stamina and type will always continue.

Right at this point it seems advisable to give a few hints as to how to tell stamina in prospective breeders. First and all important, never use breeders that have ever had any serious ailment or sickness, no matter of how good a show quality nor how well they might appear afterwards. Alertness and activity surely define stamina; full, bright eyes, healthy red head and adjuncts, and glossy plumage are its accompaniments. A vigorous male always walks on the tips of his toes, wings held poised from body, tail strong and upright; the female is a busybody every waking minute. This pair are off the roost before break of day and are the last to go in at night. Good health simply will not let them keep still. Their every appearance tells of the joy of being alive. This is stamina, which keeps them almost immune from disease and should be the first great law in these breeding problems.

Relative to color, bear this in mind: that which is in the original sire and dam which stands for color identity, when poured into the veins of the young each consecutive year can do nothing less than produce original colors of dam and sire. This is on the same principle that if equal parts of brown and white paint are mixed, the mixture then contains one half of each; continued use of brown paint will drive out the white color until the mixture shows no other color but brown. This I have proven to be an exact comparison with my breeding operations. Such methods are not according to Mendelism, Galtonism nor any other "ism" except common-senseism, but it brings results worth while.

Yet there are minor laws which affect both the wild and the domestic alike, especially among birds. These minor laws are such as appertain to certain colors and markings for male and female according to species and variety, such as the law of single lacing, of double lacing, of penciling, stippling and of spangling, as well as of certain characteristics of black, white, buff or red birds. I shall treat these minor laws under separate heads in detail, as it is here among these minor laws that the beginner loses his way.

First he must master the major laws of stamina and line-breeding; then he must take cognizance of these minor laws which I shall designate as "guide posts" to keep the beginner on the right road. The laws of single lacing and spangling are in no relation to the laws of double lacing, penciling or stippling; the laws of double lacing, penciling and stippling are related; the laws of

LAWS GOVERNING THE BREEDING

single lacing and spangling are related. Then there are the laws of red, buff, black and white breeds, and certain little sub-laws to which all are subject when perfection is the goal. These minor laws which appertain to the breeding of standard fowls, while exceedingly important, are but little understood even by skilled breeders, so called, and this is one of the reasons why so few attain success with any breed of their choice. The cut-and-fit, haphazard, guess-at-it kind are all too prevalent. So also are the short cuts for quick results which are like the get-rich-quick schemes and generally as disastrous to their devotees. These and other methods, or more properly the lack of methods or plans, swell the percentage of culls in all breeds far beyond that of the meritorious sort. It is because of this that I am endeavoring to set up these guide posts that the road which leads to success, so mysterious and strange to the beginner and even to many old breeders, may be clear to their sight that they may keep from straying into paths, dark and gloomy, full of stumbling blocks and pit-falls. The right road is never rosy at the best.

I have written at length of stamina and line-breeding, "the major laws." Obey them. Let the fittest survive, yet always the fittest according to the minor law which governs the variety chosen. The minor law which governs shape or type in any breed can be easily understood and followed if plenty of common sense is used in its application. Don't try to fit a round plug in a square hole. In other words, if you want to breed oblong type, don't use birds of circle curves, or vice versa. Remember that like begets like, if persisted in. Uniform shape can be secured in three years' breeding. It takes ten years to accomplish as much in color and markings, and naturally requires more patient investigation and study into seemingly unimportant details, and a rigid adherence to the laws which govern color and markings.

CHAPTER II

Black or White and Black and White Breeds

THE laws that govern fowls of a combination of black and white are similar, whether barred or mottled or segregated in certain sections. One of the greatest proofs that nature has inexorable laws that must be obeyed is this: that double mating must be resorted to wherever a man-made standard of color points runs counter and contrary to what nature requires in male and female according to breed. While double mating shows man's versatility and adaptability, it also shows his perverseness and his weakness as opposed to the mighty forces which control all nature. Since the inception of double mating nature has successfully resisted all man's efforts to produce other than make-shifts to conform to the aforesaid man-made standard, one to produce males and one to produce females according to written requirements; whereas nature's standard demands but one mating to produce standard male and female according to kind. The modus operandi of double mating only proves how eternally right is nature.

Search where you will among domestic fowls or wild birds and the results are the same. All barred birds according to nature's standard have a male in color lighter than the female, except in breeds where male is hen feathered. The Barred Rocks, around which the battles between double and single mating have raged the fiercest, have only vindicated nature's demands, namely:—the cockerel mating of this breed demands darker colored dams than sires; the pullet mating calls for the light colored sire. The accurately barred silver penciled Hamburg hen requires a mate of the extreme type of light color and both are produced and reproduced accurately according to standard from one mating. The hen feathered Campine is a replica of his mate and they reproduce accurately their kind from one mating. No advocate of single mating in Barred Rocks has been able to produce and then again reproduce a sufficient number of both sexes alike in color and markings to warrant classing such a production as a distinct breed; if Barred Rock males were hen-feathered the problem would be solved.

In this fight to thwart nature, the direct opposition to nature has been nature's best defense. And paradoxical as it may seem the only way to successfully combat nature is to yield, obey, acquiesce; learn her secrets and the vulnerable spots in her armor that she may do your bidding seemingly in opposition to first principles. This has been done and can be done again even with such a breeding enigma as the Barred Rocks, by strict observation to in and line-breeding as laid down in Chapter I of this book and as per chart, which

LAWS GOVERNING THE BREEDING

sums up this fact: that something in the tissues, blood and bone of a certain sire which stands for color identity, cannot help but reproduce itself in the majority, if blood of same sire is returned enough times in its progeny; in other words sire is bred back five years to its own young.

But even thus, nature fights with forces of a certain law of deterioration whenever in-breeding is practiced as before mentioned, and again the eternal problem of how to succeed confronts the breeder who thus dares nature at her most vulnerable point; and by this has the safe plan of line-breeding been formulated. (Which is not in-breeding in the direct sense.)

With apology to some unknown poet, I quote, "Persistency thy name is e'en more great than all the bubbles blown by fortune's breath." To produce exact purity of blood in domestic fowls is an impossibility, as proven again and again by the law of atavism and can also be proven mathematically. A few short minutes of figuring will convince the most skeptical; after leaving seven-eighths blood or eighty-seven and one-half per cent. pure blood, the figures get into the nineties and after ninety-nine it is always and forever a fraction less than one hundred or purity, a convincing explanation of stubs, side sprigs, off-colored feathers, etc.

In mottled breeds the law generally is for a predominance of black over the white. In some mottled breeds the man-made standard specifies one white feather to five black ones; but this is a specification written without consideration of existing conditions. There are over 8,000 feathers on a fowl. No poultry show requires that a judge shall count the feathers to find out the proportion of white feathers as per specifications, and no judge would do it if told to do so; therefore being obsolete by reason of its absurdity it is not a law.

An even distribution of the white, with black greatly predominating, is the main requirement. The similarity between barred and mottled breeds is the tendency of the black to spread, run or streak into white, which destroys harmony, and as like begets like and imperfections come without breeding for them, avoid using birds with such defects. The black should be black and the white white without any intermixing. In mottled breeds a disposition for adult birds to moult out white is simply nature's sign of the limit of value in strength and stamina as breeders. And while it means that vigor is on the out-tide, because of the lack of ability to secrete the black pigment, it does not mean that such a bird will produce young lighter colored than standard requirements; although again, as like begets like, such a bird will throw young with like tendencies at same age. Instances there are a plenty of individuals so vigorous among mottled birds that they would retain standard markings and color till five or six years of age. Therefore surely these are the proper birds to breed from to perpetuate this desirable quality.

Where white and black markings are segregated in certain sections as in Light Brahma markings, nature's control of the pigment appears to be under a different law wherein under-color plays the most important part in the proper segregation of black according to sections. Standard reads, "under-color white, bluish white or dark slate;" yet nature's laws show that a clear white under-color means generally imperfect primaries and hackle, either too light or too black. This last is possible as Standard requires primaries with white at edge, whereas many primaries can be found entirely black. This while not a serious defect, is a defect nevertheless and mostly found on inferior birds. Another defect found in birds of white under-color is a strong brassiness on male and creaminess of under-color in female. On the other hand, where the under-color is slate unrelieved by white, nature plainly voices its protest against too much coloring matter in under color by decidedly smutty hackles and, in males, intense heavy striping in saddles and backs, a breaking out of black surface pencilings on breast, body and fluff, peppered wing bars, many times attended by brassiness; in females black in backs, smutty, stubby hackles and pencilings on body and fluff. But almost invariably with the under-color bluish white, or with fluff next to skin white and that next to web blue or slate, the black points are standard and the white points free from penciling of black and brassiness or creaminess in under-color. Brassiness or creaminess in fowls of above markings appears to be created by unproportionate distribution of the black pigment. In white fowls the above trouble may, in most instances, be traced to breeding stock of short pedigree in which bloods evidently of color have been introduced. It is also a well known fact that some white breeds originated from sports of dark blood. The action of this dark blood has the same effect in white birds as the black pigment in black and white birds as outlined above, and cannot be eradicated until a sufficient number of years in line-breeding have elapsed to accomplish the cleaning out of the dark blood. Take for instance the White Plymouth Rock or Wyandottes of well known long pedigrees. They are pure white, stay-white birds; yet the time was not so many years ago when brassiness and creaminess in these same strains were grievous faults. The same can be said of our best strains of White Leghorns. Looking at the other side, we find much concern over a pure white, stay-white Orpington or White Langshan. Brassiness and creaminess are present-day abominations in these breeds, yet easily explainable by just this simple fact of a short pedigree, or the fact that time enough has not elapsed since they were originated to have bred out the foreign or colored bloods by selection and inbreeding. Those who would succeed must possess their souls in patience, as the successful breeders of other white breeds have done. They must strip their minds of all superstitions relative to yellow corn and such ilk, beliefs without tangible or sufficient proof as many strains of pure

LAWS GOVERNING THE BREEDING

white, stay-white birds have been fed on yellow corn for years. But it is a fact beyond dispute that even stay-white birds will throw a yellowish tinge to the plumage, and white corn or any feed which causes overfat will produce this same yellow tinge. The very fact that it can be removed by proper washing or even bleaching proves that it is on the outside of the feather and not on the inside where the supposed yellow corn color would locate if, as is believed, it came through the system. All the bleaching and washing in the world would not remove that color from the inside of the feather. There is no other way except the long pedigree through in and line breeding to make permanent this stay-white quality so much desired in white birds. To breed for white birds should be the aim, instead of to feed for white birds.

Black feathers or flecks of black called ticking are sometimes found in the whitest birds and are but the sign and seal of the law of atavism. Red feathers are indications of close relation to colored blood, just as strong brassiness shows a short pedigree. Lack of care or not enough shade in summer accentuates brassiness but does not produce it in stay-white birds. When in the choice for breeders one finds a male with creamy or yellow under-color with clear white surface, and another with brassy surface and snow white under-color, take the first mentioned in preference always. The brassy surface is bred in and can never be removed, while the creamy under-color is most always caused by the oil of the bird and can be washed out. Chicks from the best white strains when hatched are often of a mouse or slate color, which they retain until six or eight weeks of age.

To the average fancier black fowls may seem to have the same shade of black; yet to the painstaking, observant breeder of black fowls there appears a color aura overcasting each black breed, superinduced by the color of skin, legs and feet. For instance, the sheen of the plumage of a black breed with yellow skin, willow legs and yellow bottoms to feet is noticeably different in its overcasting aura from the sheen of a black breed with black or blackish legs, white or flesh colored bottoms to feet and with white skin. On the yellow skin fowl this many times is apparent in a bronzed appearance of what should be the greenish sheen. This is more tangible to the average fancier than the color aura before mentioned. The bronzing is related to the yellow skin and not necessarily to the objectionable purple barring. Purple barring bears the same relation to black fowls that brassiness does to white fowls and its permanency can be traced to the same cause, namely, short breeding and infusions of foreign blood, and not always lack of care or ill health.

Nature always asserts itself in surface indications on black fowls wherever there is trouble internally. In such cases the plumage takes on a brown dingy cast throughout. Sometimes it only appears in plumage under wing and in under-color of fluff. Such a bird should be discarded as a breeder. The brown tinge sometimes seen in summer just before and during moult should not be confounded with the above; search for this trouble in mid-winter or just before the breeding season.

One proof that in and line breeding will eradicate the evil of purple barring is found in that old Chinese breed, the Black Langshan. I have handled many hundreds of them in the last ten years. Their remarkable freedom from purple barring in any and all strains, under any and all conditions, and the persistent disfiguring purple barring so prevalent on newer breeds of black fowls lead me to these positive conclusions. The bronze overcast before mentioned, if found on a white-skin black fowl, indicates an infusion of yellow skin blood and a careful survey of the bird in question will show traces of yellow in skin and legs or bottoms of feet. White or gray feathers in flights or tails of black breeds are not so much indications of foreign blood as they are of imperfect moult, injury or lack of vitality. Lack of shade, negligence or ill health contribute to accentuate purple barring or sheen but never produce it to any great extent on black fowls of long pedigree. Red feathers found in black birds are not signs of impurity in any and all black breeds; for instance, the Black Sumatra, one of the very purest of black breeds and a very old one, produces many birds with red feathers in black plumage. The chicks of most black breeds at hatch are black and white in about equal proportion, which colors they retain until feathered.

CHAPTER III

Red and Buff Breeds

IN the breeding of domestic fowls, past performances or history of the breed plays a most important part in a clear understanding of nature's requirements for certain kinds or families. Nature decides that like must beget like, yet this does not always follow in domestic fowls because domestic fowls in general are in the crude state of evolution. But a knowledge of the foundation bloods in each breed's makeup helps in the proper selection toward the end in view; namely, like begetting like. Most red and buff breeds are traceable to the same source so far as the obscure past history of breeds of these colors shows. Unlike penciled, stippled, laced or spangled breeds, their counterparts are not found among wild birds. Some authorities claim that the despised Pit Game of today was once in a wild state. History proves their domesticity hundreds of years before the Christian Era.

The Brown Red, the brilliant ginger Red Pit Game and the massive Malay Game seem to be the only fowls known of ancient lineage which are nearest to completeness in red coloring throughout, as are our red breeds of today. That Pit Game blood has been used in the production of many of our standard breeds is a well authenticated fact. The similarity of color laws in ginger red Pit Games and our newer red breeds shows the relationship and the foundation of their brilliant ruddy hues and I may go still farther and call attention to their hardiness, enduring qualities and prolificacy as derived from the same source. The power of proper selection and in and line breeding is clearly shown in the evolution of the red breeds as seen today, by man-made standards in which nature's laws are seemingly ignored. All ancient breeds of red plumage, whether wild or domestic, have slate under-color, yet in one breed of red fowls (which one writer describes as a conglomeration of mixed up indifferences) the under-color in a vast majority has been purged of all slate color, yet retaining black in wings and tail, at the same time toning the surface color into desirable evenness to correspond with under-color. Yet many times the purging has been too complete, in that wing flights and tail coverts show absence of the necessary black. Thus does nature tell the red breeder to retrace his steps. By the same process which produced the first can the second be accomplished; namely by in and line breeding with proper selection. On the other hand, by partly maintaining the slate in under-color, another breed of red fowls has been made to take on a characteristic red surface all its own; a hue which while apparently of the same color, when placed alongside the first mentioned presents an altogether different red color aura readily distinguishable even by a novice. This too has been accomplished by that first great law to which "like begets like" owes its existence.

Like all other breeds the red varieties are subject to the minor and sub laws which help or hinder as the case may be, according to their application or non-application. The minor laws which govern the color schemes of red or buff fowls are more or less empirical in the nature of deductions as to the whys and wherefores and are based entirely on the results of many experiments with these allied colors. The process is curiously like the mixing of paints; it requires an artistic skill in the mixing but demands unlimited patience and perseverance to get the correct shades permanently fixed and sustained.

The first experiment made to bring about the standard red and buff shades was to mate a bird of dark shade to one of light shade. Many repetitions showed the futility and I may say foolishness of this experiment. The result in the first cross simply showed in one bird the color characteristics of both sire and dam intermingling in a motley array of dark and light colored feathers, far from the ideal, even surface of standard description. Then came the idea of grading colors or, by careful selection, gradually smoothing in and blending the shades into the one desired. This is practiced today by all successful breeders of red or buff fowls. It is not accomplished in one year or even two. It requires deep study, intense application and patience. This method of blending or grading colors is applicable to either red or buff fowls.

The first procedure is to take a male of uniform even surface throughout and match his breast and body color to the surface color of the female, said female to be of even surface color throughout. This often brings about most excellent results in the first cross and a continuance in this manner by line breeding blends and fixes the color desired. But there are many points to be considered even thus, which may be classed as sub-laws, and a knowledge of the same is of great assistance in acquiring a purity of color throughout. Purity of color means in nearly every instance purity of blood. In red fowls the under-color, while secondary in the show room, is nevertheless a potent factor in the breeding pen. An even, rich, brilliant red surface is enhanced and supported by a deep red under-color and in the breeding pen has the power of transmitting and strengthening the color in the young. Permanent slate or smoke in the under-color deepens the red surface to the color of mahogany bay. Slate or smoke following feather to skin has a tendency to dull lustre of surface coloring. Slate at top of under-color next to surface with red at bottom next to skin is not permanent in chicks if found under a brilliant, even red surface, but is permanent under an even mahogany bay surface. These laws of color effect seem nullified where surface is of uneven color or one of various shades. Brown harmonizes with dark chocolate surface and both are on the downward slant beyond the highest or most brilliant point in red color tone, as the dulled aura shows.

A light red or silvery red under-color is very unstable and many times in second moult degenerates into white patches throughout undercolor, and is coupled generally with weak changeable surface color. Too dark surface color without slate or smoke underneath is generally accompanied by the dangerous silvery red undercolor. Surface color with a silvery red or roan overcast is allied to the aforesaid silvery red under-color and is a pernicious defect hard to eradicate. Black or brown stripes in hackle of young males never moult out, but black tips at base of hackle will moult out. A male which shows these black tips in chick feathers will not have them in adult plumage; but this marking in chick plumage indicates a future breeder of pullets with the required black tips in hackle. Old males sometimes show this black tip in hackle when part through the moult but when moult is complete it is not there. Males with the permanent black or brown stripes in hackles will sire pullets with an overplus of black in hackles and if said males have a light shaft in this black or brown stripe, the females from such a sire will invariably be covered with these disfiguring light shaftings on entire body, back and breast plumage. Young males will also carry the same on body and breast surface. Lacings on edge of feathers show the union of two birds a few degrees apart in color tone and is related to mottling, which comes from extreme matings.

Golden hackles, or "pumpkin necks," as they are derisively called, mar an otherwise even surface and are nothing less than the result of bad breeding and careless selection. All red breeds should carry more or less black in flights and secondaries and in main tail feathers. Buff breeds should be free from black in any part of their plumage; yet both red and buff breeds carry a mixture of yellow and red pigment controlled and subdued by in and line breeding according to kind. Black has an affinity for red and is necessary in toning to shade of red desired. White has no affinity for red; hence is of harm in any part of the plumage and as aforementioned bears a relation to silvery red under-color or surface overcast with roan splashes.

At times in the buff breeds, the red pigment is stubborn and proclaims itself in surface of secondaries and in tail by a chestnut hue which is the despair of the buff breeders as it spoils harmony of the golden buff in the other sections. The only explanation of its persistence is that the larger heavier feathers draw out an over-plus of red pigment which seems to bear a relationship to the black points in red breeds. Its removal calls for the thorough cleansing power of persistent line breeding and intelligent selection in matings. Find a sire and dam free from these defects and then line breed according to chart. As before mentioned, wide contrasts in color of sire and dam produce mottled or patchy surface color. As stated in red breeds, anent hackles with brown stripes and the light shaft, buff sires with this defect breed chicks of both sexes with the objectionable light shafting throughout the entire surface plumage, and also produce females with hackles three or four shades darker than body color. Under-color in buff breeds does not play so prominent a part in the control of surface color as it does in red breeds as it seems to have lost its power of control in toning down from red to buff. This is substantiated by the fact that a deep buff under-color generally keeps company with a reddish buff surface; and also by the fact that too light or almost white under-color is covered by a lemon shade. In this last, lack of pigment is prone to follow into flights and main tail of both sexes.

In Buffs, the best strains produce chicks at hatch of a soft creamy buff, many times with darker shades of buff on back and head. In Reds, long pedigreed parents give chicks at hatch a strong, even red shade. Where silvery red under-color is in sire or dam, or both, the chicks are of a light primrose color with occasional dark patches on back and head which are present until chicks are feathered.

CHAPTER IV

Penciled, Double Laced and Stippled Breeds

IN the penciled and stippled varieties of fancy fowls there is a decided similarity in the laws which govern them. This similarity leads the investigator to the belief in one original source. They are therefore related, as their many similar characteristics prove, yet with sub-laws controlling the differences between penciling and stippling. The main law which proves their relationship is the fact that the males of all penciled and stippled breeds or varieties must have black breasts and bodies. This is an imperative law as regards these breeds and cannot be transgressed if one expects favorable results. To illustrate: a certain well known fancier and friend conceived the idea of a penciled Brown Leghorn with male penciled in every section like female. The result of six years of effort shows the impossibility, as every male comes with black breast and body. I am absolutely positive that he can breed till the crack of doom and not produce a penciled breed with males having penciled breasts and bodies. He is working contrary to the laws governing penciled or stippled breeds. Yet in seeming disregard to all this are the two exceptions, Dark Cornish and Red Caps. Both breeds show males having solid black breasts and bodies, and females of Red Caps not penciled or stippled but with an abortive or crescent shaped spangle. Females of the Dark Cornish have both lacings and pencilings. A lacing goes around edge of feather; penciling is on the feather away from edge. A Dark Cornish female has a lacing and also a penciling, the only known breed so marked. The males of the Dark Cornish have black breasts and bodies because of the strong influence of two of the breeds comprising its makeup, namely, Black Sumatra and Black Breasted Red Game. This influence partly controls the tendency to single lacing in the females. Single laced females in this breed will produce males with splashed breasts and bodies, again showing how positive are the laws of nature as regards markings according to kind. While authorities are silent as to the origin of the Red Caps, the feather characteristics show breeding of similar character to the Dark Cornish. They indicate strong infusions of black blood and black breasted red blood, undoubtedly with Golden Hamburg blood, which would account for the black breast on male and the abortive spangle on female. But barring these two exceptions, the rule is that black breasted males do not belong to any species of fowls or birds carrying lacings or spangles and only belong to those species carrying pencilings or stipplings. All penciled breeds are alike in distinctive markings. All stippled breeds are alike in distinctive markings. While male of penciled varieties is similar to male of stippled varieties in having black breast and body, the similarity ends there, except where double mating is used. This only proves their relationship as well as the sub-laws which govern each kind and breed. In every breed or variety where female is penciled the male carries a black stripe in hackle and saddle. In every breed or variety where female is stippled the male does not carry a black stripe in hackle and saddle except in brown and silver leghorns. Standard description.

These two varieties call for black stripe in hackle and saddle which cannot be produced without double mating. Double mating in this instance seems to break the sub-laws governing penciling and stippling and yet it proves those laws, as it is a well known fact that females of the male line of these two breeds are more inclined to penciling than to stippling; that males of the female line are without black stripe in hackle and saddle or very deficient therein.

Another peculiar fact in connection with these sub-laws which, while not exactly pertinent to the subject, is well worthy of notice, is the phenomenon of color harmony or color aura. Every male of every variety carrying red plumage with black stripe in hackle and saddle should have yellow legs to perfect the color harmony or aura; which accounts for the exceeding beauty of the standard bred Brown Leghorn male. Take the same bird and remove the black stripes and a dirty rusty red plumage is the result. Take the same bird again and put willow, green, white or blue on him, as he is without black stripes in hackle and saddle, again the color aura is perfect and a beautiful bird appears. Put in the black stripes with the willow legs, etc., and a coarse harsh color greets the eye. Seemingly in support of the above, most stippled breeds have willow legs, etc., most penciled breeds yellow legs.

In stippled breeds it is a law that the female shall have salmon breasts, yet some breeders of silver varieties that are stippled, claim they should have silver breasts as they claim it enhances the entire silvery grey plumage. Yet silver breasts are related to pencilings not stipplings, and absence of salmon will cause pencilings on breasts and also tends to produce pencilings or coarse stipplings on back and wings. This shows conclusively a transgression of law. Furthermore, such females will have a tendency to produce black stripes in males. In stippled breeds, males having an inclination to dark stripe in hackle, with shaft of feather light colored, will produce females full of that objectionable shafting on back and wings, as well as coarse stippling. This light shaft in hackle of males of penciled breeds and even laced breeds is the source of so much light shafting on back and wing of these breeds. It is a very bad defect in any breed and it is found in nearly every breed even to white and black birds. It should be avoided in every case in the breeding yard. Note that its source is in

LAWS GOVERNING THE BREEDING

the hackle of the male and can be avoided if proper care is taken in selecting breeding males free from this defect.

Shafting, brickiness and pencilings are closely related, because all males of penciled breeds have a black stripe in hackle where shafting springs from. Brickiness is allied to the red plumage of many penciled breeds and again proves broken laws when these defects are found in stippled breeds; therefore the deductions are that a breeding male of a stippled variety must have no light shaft to hackle and must have a solid black breast; that the females must be free from inclinations to pencilings, shafting or brick and have clean salmon breasts. The above also applies to males of penciled varieties, especially as regards black breast, as splashed breasts on breeding males mean an inclination to lacings on female young from such a sire; another indication of a broken law of infusion of blood foreign to kind.

In penciled females avoid using those with irregular pencilings or those with bars across feathers. This shows poor selection as well as haphazard work in breeding. Such females incline to produce males with smutty hackles, shoulders, and saddles. In choosing breeding females of either penciled or stippled varieties, see that small feathers covering the entire under side of wings and small feathers inside of tail are accurately penciled or finely stippled according to breed. Some breeders choose their males by the same method but choose them when only eight weeks old because then their chick feathers show pencilings or stipplings on entire surface and their relative breeding value. Chicks from penciled, stippled or double laced breeds are hatched with the so-called chipmunk markings or regular stripes of dark and light color and substantiate the claim of one original source for these breeds.

CHAPTER V

Single Laced and Spangled Breeds

THE history of single laced and spangled breeds shows a common source and origin. Many characteristics which are brought out in breeding plainly prove this kinship, wherein the laws that control the marking of single lacing and spangling are in strong harmony one with the other and can be evolved into lacings or spanglings according to the desire of the breeder. This calls simply for the sub-law of right and accurate selection as to markings wanted. Investigations and observations of pheasants and other wild birds further prove this harmony between lacing and spangling, and the accuracy of these markings true to kind in wild birds shows the complete obedience of nature to her own laws which are immutable and unchangeable. The breeder who desires to advance must never lose sight of this fact and obey implicitly if success is desired.

I have said that the mysteries which surround the breeding of high class fowls are not laws, but the results of broken laws; i. e. the mingling of breeds of certain and fixed characteristics foreign and hostile one to the other and which in markings show no relationship nor harmony in perfect alignment of nature's law that "like begets like." With the progressive and true fancier, to know is to obey. To start right, I must reiterate, is nearly as important as keeping right. The first thought in starting must be to choose birds which embody to a high degree stamina, type, color and markings according to kind and standard. Then in-breed and line-breed, obeying the minor laws and select each year for breeding only those stamina carrying the most apparent characteristics of kind. Nature has placed these characteristics plainly where all may read and learn. Strictly observe and obey these minor laws and the puzzle of mating and breeding will be solved and one can do nothing less than produce a majority of high class show birds with something back of them to continue in the right path.

It is a law of single lacing that the female shall be laced in every section and in harmony with the markings of the female the male must also be laced in every section. Although the long feathers of saddle and hackle of males in most laced breeds have a tendency toward a rayed appearance, yet it really is a lacing when shape of feather is considered. This harmony in the lacings of male and female can be seen in all its perfection in the Sebright bantams. All varieties or breeds bearing single lacings are subject to the same immutable laws whether major, minor or sub-laws.

In choosing breeding birds, and, by the way,

choose only show birds of high class, there are sub-laws which should be observed and obeyed. For instance, a single laced male or female of any variety, breed or kind, with all the small feathers covering the entire under side of wing, well and clearly laced, and small soft feathers inside and back of tail, well and clearly laced, is always a breeder par excellence of well laced cockerels and pullets. Provided, however, that lacings on surface are as they should be. This is an unfailing guide in any parti-colored breed. Narrow lacings should always be selected in breeders. A narrow laced pullet will invariably be free from mossiness in any section and more valuable as a breeder as well as a show bird, because when she moults as a hen she will still hold her color and markings free from mossiness. Mossiness in single laced birds is one sign of penciled or hostile blood and by selecting always the above narrow lacings when choosing breeders the eradication of this defect can be accomplished.

As regards males, choose always those with complete laced breasts and bodies free from any tendency to black. Black in breasts and bodies of males is also an indication of the taint of foreign infusion. Females with mossiness in plumage produce males with breasts and bodies inclined to black; males with black in breasts and bodies produce females with lacings spoiled by the disfiguring mossiness throughout plumage. In short, obey the law of single lacing as before stated. Mossy lacings and black breasts are pitfalls and stumbling blocks in the road of progress to high class breeding of laced fowls. Weed out also the birds with ragged and uneven lacings; those carrying lacings with almost square ends, which have a resemblance to half spangle and half lacing. The true lacing is narrow and follows perfectly the shape of feather from under-color to under-color in every section.

A well laced male with smut in shoulders and saddle is a poor breeder, although generally these defects go with black breasts, poor lacings and deficient wing bar and are further indicatons of foreign blood and broken laws. In spangled varieties the relationship to lacing is shown by the tendency for one to revert to the other and also shown by nature's minor law which requires that male and female shall be spangled in every section alike, excepting, of course, the shape and nature of male hackle and saddle feathers, although there are spangle breeds wherein many henny feathered males are found which are exactly like the females in every section, just as the males of Sebright bantams among single laced breeds are exactly like the females in lacings in every section. This further proves the kinship of nature's laws in lacings and spanglings. One further proof still, is that there has never yet been discovered a complete henny feathered male among penciled or stippled breeds, which same have no kin in markings or feather characteristics with laced or spangled breeds. What has been said of single lacings relative to guides in choosing breeders,

LAWS GOVERNING THE BREEDING

etc., and avoiding pitfalls and the like can be applied to spangled breeds, always considering the direct difference between a lacing and a spangle. A lacing goes around the edge of feather; a true spangle covers the tip of feather and comes to quill in shape like the tail of an arrow, or V shaped. In spangled breeds, imperfect spangles go with peppered flights and tails and ragged and uneven marked shoulders. A common defect is the spangle having the appearance of an abortive lacing. Sometimes overdone spangles give the bird the appearance of black in breast and back; such a bird will carry generally smutty flights and main tails. Abortive lacings or poor spangles show bad judgment in selection and are a sign of the careless breeder. Overdone spangles show an infusion of black blood; neither are necessary when there is absolute obedience to the laws of line-breeding.

In spangled and laced breeds the chicks are hatched with dark and light colors, irregularly and indistinctly blended from one to the other. In making up the breeding yards of any and all breeds it is well to take careful note of what the standard says relative to the most typical specimens, and those nearest to standard weights, where weights are applied, being most useful. Surely from a fancier's standpoint and also a breeder's standpoint, they are not abnormal in any way and should embody stamina and good health to the highest degree.

CHAPTER VI

The Breeders and Fanciers

THE laws of breeding standard fowls contained in this book are written especially for the beginner, not for the old soldier who has fought these things out on the same battle ground as "yours truly." The term beginner covers two classes. Contrary to accepted rules of speech, poultry culture has "new beginners" and "old beginners." The "new beginner" or novice is indeed quite new and is very susceptible, without discrimination, to every bit of advice and counsel offered; yet withal one who progresses with astonishing speed until hoisted on the petard of his own conceit. This same conceit is a peculiar trait of the "new beginner," a conceit which keeps his head high in the clouds of his new found knowledge, hiding from his vision the stumbling blocks which bring him again to earth and common sense. This conceit is attributable to the fact that the novice is generally very successful the first year; the second year finds him wiser than he can ever afterwards hope to be. After the novice passes this teething stage he becomes a breeder worth while and ranks with the philosophers who believe in the progress of the future.

LAWS GOVERNING THE BREEDING

But the "old beginner," he is the original drag on the wheels of progress; worse even than the fossils who live in the cobwebs of the past. The "old beginner" in poultry culture is the literal definition of the term, always beginning and never advancing, whose claim to tenacity of purpose is that in spite of repeated failures of theories and plans, he is still an enthusiastic poultryman, growing older in years with a greater accumulation of theories but proving nothing. This "old beginner" finds one year that the penciled breeds are the best ever; the next year the laced breeds are in his favor; then barred breeds attract him; blacks or whites, reds or buffs and so on. He is never long enough with any one breed to be other than what his name implies; an "old beginner." He is ever the slave of capricious whims, scorning advice and proven facts. Yet to him I offer these pearls which have been the backbone of every successful fancier and breeder for many decades past and will be for many decades to come. These laws are not intended by the simple reading of same to turn out skilled breeders and fanciers. Faith without works is dead; they must be practiced and applied, studied and tried. A complete realization can only come by years of actual experience.

The thoughts written here can only serve as hints and warnings, rudimentary at best, yet based entirely on a life-time of investigation and observation of nature's laws as applied to domestic fowls. I have such absolute confidence in the laws of in and line breeding as shown on chart on page 15, that I feel capable of taking any fowl, dog, cat or any other short lived animal of any color or markings and producing in ten years, one hundred of similar markings and characteristics. Note that the heart, nature's great pump, has exactly the same method in sending the blood to the tiny toes of the humming bird as it has in sending the blood to the trunk of the mammoth elephant. Just as exactly is all animal life governed to the same results by in and line breeding properly applied.

Perseverance and patience are the handmaidens of the successful breeder. Jumping at conclusions and impatient desire for quick results work the ruin of the average fancier. He does not differentiate between the perplexities attendant on the mingling of bloods and clear results of nature's pure breeds. That very impurity which produces such a variety of culls and mediocre stock confounds, by the unexpected production of a high class world beater, either male or female, therefore the average breeder jumps to the conclusion that double mating is the thing. He does not realize that this phenomenon goes no farther than the male or female that produced it, unless years of in and line breeding are practiced with this same phenomenon as the foundation. This last the average breeder has not the patience to do and failure is the result so far as the reproduction of kind is concerned.

Without the perplexities and puzzles no skill

LAWS GOVERNING THE BREEDING

would be required in these breeding operations. There would be no incentive to higher achievement; simply a drifting, purposeless existence. The homely saying that any dead fish can float down stream but it takes a live one to swim up, is most applicable. The fanciers of indomitable purpose and the courage of their convictions, and with the never say die spirit are the ones who have solved the mysteries of animal breeding and brought into existence animals of the highest type for all purposes for the pleasure and benefit of mankind.

What one man has done another, with the same fixity of purpose and determination, can do. Mr. "New-beginner" this means you; for with you there is hope. Mr. "Old-beginner," whose greatest success is in making failures, take heed, and incline thine ear to goodly counsel.

And now a few words relative to the vital points brought out in the foregoing chapters. First, in regard to the possibility of deterioration from in-breeding. Every successful breeder believes there is a deterioration. Most authorities claim this to be a fact. Yet modern scientists and theorists believe to the contrary, and hold that all that is necessary is to select for stamina and vigor, no matter how closely related the breeding stock. My advice, after forty years' experience, to every breeder is a quotation of an old railroad rule which reads: "In case of doubt take the side of safety." The salient successes that mark all history in all animal breeding have shown that the side of safety in animal breeding is in believing that there is a deterioration.

Speaking for myself, one experiment along these lines embracing a period of four years proved to me most conclusively that deterioration from in-breeding is a fact. I was breeding Pit Games at the time. Cockers of those days and even today in eastern New York and western Connecticut practiced out-breeding always to perpetuate stamina, vigor and fighting qualities. They never used a male in breeding anyway related to the females. Color and markings were ignored. The only qualification was to stand the gaff. In-breeding was believed to lower vitality and to produce inferior birds. Yet contrary to this, Harrison Weir's book quotes many instances of grand fighting strains long and intensely inbred. So mote it be. Suffice to say my experience was vastly different. Desiring to perfect feather markings and color from a beautiful pair of Pit Games so that the progeny would be of one color and markings, I in-bred both sire and dam to their young for four seasons. I was careful in the building of yards to have them covered to prevent contamination from my birds flying out or others flying in, and was as watchful in that respect as every first class breeder of Pit Games must be to preserve their one essential quality. Yet at the end of four years I had a dandy strain of runaways; feather bred in but courage bred out. This to my mind was nothing less than a deterioration from in-breeding. Believing thus,

LAWS GOVERNING THE BREEDING

I have always practiced out-breeding while in-breeding. Or putting it more clear, after line is established by in-breeding, as per chart, I choose all breeders from the farthest removed in relationship, thereby practicing to a great extent virtual out-breeding, or in the modern term, line breeding which means preserving stamina and vigor while keeping same blood line pure.

As regards double mating, it is a valuable commercial short cut, but an uncertainty and a delusion to the uninitiated. Line bred birds produce show birds from one mating, of which there is proof galore. One sub-law worthy of reiteration is that one respecting the light shafting in hackles of all males whether black, red, buff, penciled or laced. It will unfailingly produce young with this objectionable feature prominent and predominant throughout the entire plumage, marring the harmony of color so desirable. Color harmony is entirety of marking and color blending into an aura free from disfiguring blemishes as so accounted in the standards prescribed for each breed. For instance, laced feathers free from mossiness, penciled feathers not barred or broken in pencilings, red not overcast with grey, lacings or shaftings, stipplings even and unbroken, devoid of brickiness or shafting, black with absence of purple, white clear of brassiness. As to yellow corn and white birds, my views are expressed in a foregoing chapter, yet the word I would leave with the breeders who feed white corn to white birds is this: travel on the bridge that carries you safe over. And now a few words of advice in conclusion. Compare points relative, spangled and laced, penciled and stippled; breed and note particularly the entire lack of relationship and the peculiar and lasting harm incident to the mingling of these bloods. Spangled and laced breeds have one common origin, yet each have sub laws which govern feather markings, but are entirely foreign to penciled and stippled breeds. Penciled and stippled breeds show strong relative characteristics, but both are governed by their own sub-laws.

Investigate the law of atavism and its relative bearing on your breeding operations; it will help weed out the culls and eventually bring a strain to a near pureness of blood, after blood lines are established. To produce a high-class strain of show birds that will uphold the law, the law that "like begets like" should be the aim of every breeder or fancier, rather than to produce occasionally, one phenomenally high class show bird by hit or miss, haphazard methods. One is the house built on a rock; the other a house built on shifting sands.

One more last thought. No breed of fowls, however full of stamina and made so by selection, can so continue unless proper common sense methods are used in the feeding of breeding stock. The greatest factor is free range for breeding stock and growing chicks, which should be fed nothing but hard grains. Never use wet or dry mashes nor any condimentals nor meat feeds, even in confinement, to breeding stock or any food whatsoever that would cause increased egg production during the breeding season. In confinement, feed abundance of green food and then some and no feed but hard grains. Excessive egg producing foods make weak eggs; therefore poor hatchibility and weak chicks. Egg factories should never be the breeding pens. The appalling death rate among little chicks of all breeds at the present time in poultry history is but the result of long years of careless confinement, ignorance in feeding and blind breeding operations.

To the careless breeder, this book is as so much waste paper. The painstaking seeker after the truth never becomes too wise nor too old to learn. Theories are visions, facts are concrete. Stamina is the first and main law from the beginning to the end. Judicious in and line breeding will accomplish results worth while. Above all keep bright the thought that those methods which made success must be persevered in to keep success.

THE END.

AUTHOR'S NOTE: The double-mating system being a law unto itself and not in harmony with the natural laws governing the breeding of standard fowls, no mention is herein made of the laws or rules governing double-mating.

List of Prominent Breeders of White Laced Red Cornish Fowls Originated By W. H. Card

Frank C. Burbank, Sandwich, Mass.

J. Thomas Harp, Buffalo, N. Y.

Geo. W. Webb, 23 Weldon St., Rochester, N. Y.

Judge H. B. May, 692 Huntington Ave., Boston, Mass.

La Vergne Dimock, Stafford Springs, Conn.

J. Grant Griswold, 591 Blue Hills Ave., Hartford, Conn.

Mrs. L. S. Judd, Chapinville, Conn.

Geo. W. Tryon, North Stonington, Conn.

Dr. G. A. Perley, Woodstock, Conn.

H. L. Knowlton, Concord, N. H.

W. L. Loope, Millerton, N. Y.

Ewell Gay, care of Insurance Co. of North America, Atlanta, Ga.

Wm. Taylor, Chief of Police, Montgomery, Ala.

Fred W. Rogers, Montello Sta., Brockton, Mass.

W. S. Templeton, Morgan Hill, Cal.

Feliciano Ferreira de Moracs, Campinas E. De S. Paulo, Brazil.

M. K. Thomas, 231 South Main Street, Wallingford, Conn.

Edward Parker Sands, Fullerton, California, P. O. Box 289.

J. O. Snyder, Yoe, Penn.

Printed in the USA
CPSIA information can be obtained
at www.ICGtesting.com
LVHW080155170524
780574LV00009B/415